Russian
T-34
Battle Tank

Horst Scheibert

SCHIFFER MILITARY HISTORY
Atglen, PA

Front Cover Artwork by Steve Ferguson, Colorado Springs, CO
Additional Research by Russell Mueller

CREST OF THE STEEL WAVE

In the smokey sunset upon the steppes east of Kursk, Russia, a Model 1943 T-34/76 of the Russian Guards 5th Tank Army prowls amongst the burning hulks of the defeated German 2nd SS Panzerkorps.

Introduced in the 1942 winter campaign at the Battle of Stalingrad, the new diesel powered T-34 had displayed great ruggedness, particularly in the long range shootouts against the infamous German "88" cannon-armed Panzers. In the great battle of the Kursk Salient in July 1943, waves of the dreaded 76mm cannon-armed T-34 "Panzer-killers" gallantly pushed to within point-blank range in a bold counterattack against the less maneuverable, heavier Wehrmacht machines. The ensuing bloodbath initiated the complete collapse of the Eastern front as the devastated Wehrmacht reeled westward in retreat. The Red Army's T-34 equipped battalions fought on for two years until the culminating battle of Berlin. As a fitting symbol of Soviet victory, T-34/84 equipped regiments would be the first Allied troops to enter the fallen Reich capitol.

Simple in design and sleek by armored vehicle standards, the sturdy "tridsatchetverka" (34) tank became the consumate symbol of the Red Army's Great Patriotic War. Easily modified and adaptable to changing conditions on the battlefield, Russian steel mills pressed more T-34s into combat from 1943 to 1945 than the total number produced by both England and Germany. Following World War II, the Soviet Union equipped many of her satellite republics and allies with the T-34, and would eventually fight again in Hungary, Korea, the Middle East and Indo-China. And, after more than forty years since its debut on the snowy battlefields at Stalingrad, late model "34"s were still in reserve service as the once mighty Soviet Union dissolved and passed into history.

PHOTO CREDITS
Bundesarchiv (BA)
Wolfgang Schneider
Hans-Joachim Schroeder
Horst Riebenstahl
Imperial War Museum (sketches)
Tamiya (sketches)
Podzun-Pallas Publishers Archives
Photo Archive No.3 (Hentschel Collection)
Tank School, Bovington, England

Translated from the German by Edward Force.

Copyright © 1992 by Schiffer Publishing Ltd.

Printed in the United States of America.
ISBN: 0-88740-405-7

This title was originally published under the title,
Der Russische Kampfwagen T-34,
by Podzun-Pallas Verlag, Friedberg.

We are interested in hearing from authors with book ideas on related topics. We are also looking for good photographs in the military history area. We will copy your photos and credit you should your materials be used in a future Schiffer project.

Published by Schiffer Publishing, Ltd.
77 Lower Valley Road
Atglen, PA 19310
Please write for a free catalog.
This book may be purchased from the publisher.
Please include $2.95 postage.
Try your bookstore first.

We are interested in hearing from authors
with book ideas on related subjects.

A T-34/76 column in the first year of the war. Signaling was still done with flags.

Development

It is not true that the T-34 — a great surprise for the German soldiers in World War II — arose out of the blue. There was simply a lack of reconnaissance.

The BT (Bystrochodya Tank = Fast Tank) 1-7 series, which originated in 1931, and the BT-IS (Ispitatelniy = Researcher) developed in 1936 featured from the start the Christie running gear with the big road wheels, which was later typical of the T-34 as well. Since the BT-7 they were equipped with robust Diesel engines, and the BT-IS was the first to have angled armor plate. All of them, including the two successor models, the A-20 and A-30, were still armed with a 4.5 cm tank gun and could travel on their road wheels even without tracks — hence the "Fast Tank" designation.

There was much in them that pointed the way to the T-34. Even more similar was the T-32 (1939) based on the A-30, which did not go into series production but was a decisive transitional step to the ensuing T-34. This design already included the 7.62 cm gun, the fifth (previously there had been only four) pair of road wheels, and it had finally dispensed with the complicated wheel-and-chain system.

Upper right: A BT-7 — its Christie running gear with the big road wheels can be recognized easily.

Right: The BT-7 could also travel without tracks. In this case, as shown, the tracks are attached to the sides of the hull.

After minor modifications, the first two prototypes of a successor to the T-32 were built in 1940, designated T-34, and tested successfully. In March further examples were tested by the troops. After improvements to the transmission, this T-34 was fitted with an improved 7.62 cm (L/30.5) tank gun (M-1938) and went into production.

This battle tank was not — as one reads so often — used in the Russo-Finnish War. However, when Operation "Barbarossa" began, 1300 of them were ready to use, even though some of them were fitted with older gasoline engines because of production shortages.

A variety of improvements — from the T-34/A to T-34/F (also called T-34/76 A-F to differentiate them from the T-34/85 after its introduction) — and the T-43, which did not go into production, led to the T-34/85. which differed from the 76 type in having a larger cast turret with a bigger (8.5 cm) and longer (L/51.5) gun. There were many versions of it as well.

The T-34 chassis was used with numerous bodies for special tasks (bridgelaying, mine removal, roadbuilding, etc.). In addition, it was used for the self-propelled SU 85, 100 and 122 guns.

Upper right: The BT-IS shows for the first time the angled surfaces of the hull — at a time (1936) when no other tankbuilding nation thought of it!

The further development of the BT-IS to the A-20 (shown here) and A-30 show an improved turret and a somewhat higher hull. In these features they resemble the later T-34.

Above: Still more like the T-34 was the T-32 developed in 1939 — which already carries the 7.62 cm gun (though with a shorter barrel). In addition, it already has five road wheels on each side and has finally given up the technically complex wheel-and-chain system.
The illustration at upper right shows its rear view.

Right: In the same year (1939), the T-34 went into production, fitted with a longer 7.62 cm gun than the T-32.

3852.41

The T-34/76, Types A to F

Originally it was just called T-34, but after the introduction of the 8.5 cm tank gun, all T-34's still in service with the older 7.62 cm gun were designated T-34/76 to differentiate them. Of it there were (according to western nomenclature) the types A through F, of which B, C and D were built in greatest numbers.

It is not easy to tell the different types apart; in fact, it is sometimes impossible, since rearmament, repairs, component shortages and production in small, sometimes make-shift factories resulted in many mixed versions — especially in terms of the turrets — than thoroughbred types.

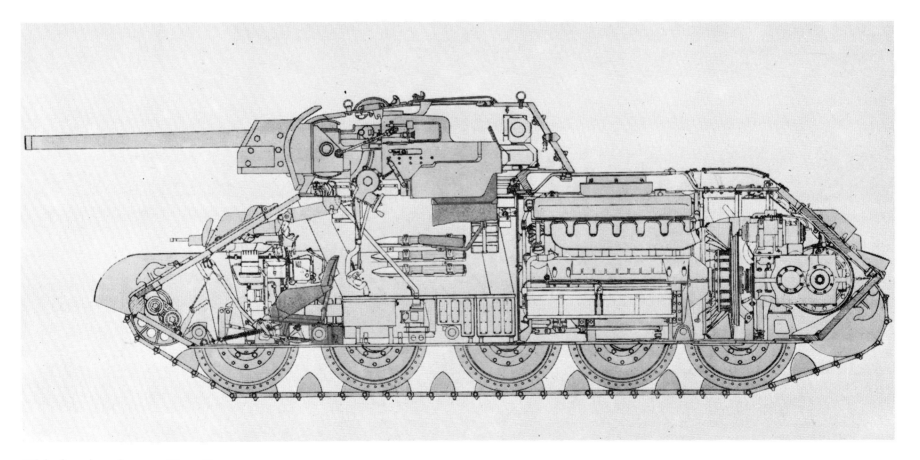

This drawing shows a Type B.

Type A, 1940

It can be recognized by the shorter (L/30.5) gun, a mount shaped like a pig's head, and the smaller cast or welded turret without a commander's cupola. The first 150 vehicles still had a machine gun in a ball mantelet at the rear of the turret.

Upper right: A T-34(A). The cast turret with the pig's-head gun mount is easy to see. (Hentschel Collection)

Right: This T-34(A) also shows the typical gun mount of the A type. But its turret is welded. The short gun (L/30.5), compared to the later type, can also be seen. It served as a pathfinder — here for the 11th Armored Division, whose symbol can be seen on the turret.

A T-34(A) with a cast turret has gotten stuck and thrown its right track. Along with the practical shape, the stronger and — compared to the German tanks of the time — longer gun (with higher muzzle velocity), it featured robust and durable Diesel motors, wide tracks (for better off-road travel) and easy maintenance. (BA)

Type B, 1941

Along with the longer (L/41.2) gun, the new angular gun cradle can be seen. But the gun mount used on the A type was also used. This version also used either cast or welded turrets.

Later versions also had waffle-like reinforcing plates riveted on. A few vehicles had antennas.

Upper right and right: Type B can be recognized by the box-shaped riveted gun-cradle protector, the longer gun barrel (L/41.2), and, at least on the later vehicles of this type, the new pierced road wheels. Greater numbers of this type were made than of type A. It could be told from the similar C type by the big, heavy one-piece turret hatch. The C had two smaller ones on the turret roof. (2 x BA)

It also has the pierced road wheels. Its turret cover, though, shows no outward bulge and round perforation.

Above: A destroyed barrel. And yet another turret cover (round hole) can be seen here.

Upper right: And here again, an outward bulge in the turret cover can be seen.

Right: "Close combat" for the tanks. Here a T-34 takes on a Panzer II — unequal opponents.

All **B** types on this page still have the older road wheels without perforations. There were still mixed types until 1944 — not only in terms of road wheels, but also varying machine guns and tracks.

11

All three photos show welded turrets. They can be recognized as B types by the box-shaped gun mount, the one-piece turret hatch and the platelike track links.

The turret hatch on the model at left is interesting. It shows a different shape than the previous models (including the A-type turret hatch cover). Here too, all the road wheels lack perforations. Note too the boxes and spare links on the hull. (1 x BA)

Above: It is hard to tell exactly how this situation came about. To judge by the plate-type tracks, they are B types. Note the mixed types of road wheels.

Upper right: A B type with extra armor welded on — in individual plates on the bow and turret.

Right: A T-34/76 (B) (note the big turret hatch cover) with the new perforated tracks. The turret is set in the six o'clock position.

Here too is a B type (note the old machine-gun mantelet and driver's window). An interesting addition to this vehicle is the second periscope seen on the right side of the turret roof.

Type C, 1942

Its cast turret now has two visors (instead of the previous single large one) on the turret roof. Instead of the earlier plate-link tracks, it has ribbed and perforated tracks. The driver's window has protected visors. The bow machine gun is mounted in a new ball mantelet.

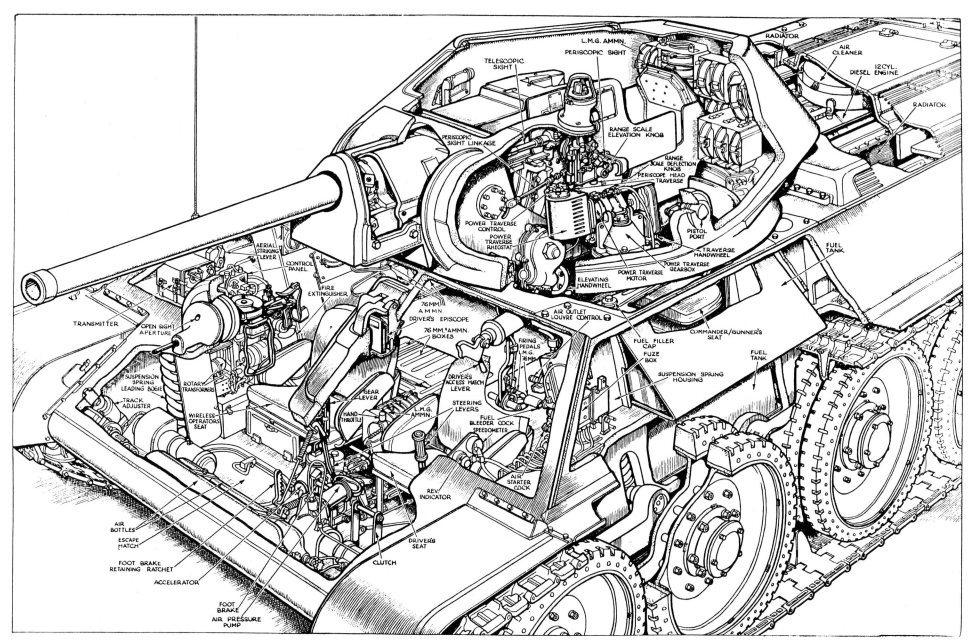

This drawing shows a C type.

Type D, 1942

Recognizable by the obvious hexagonal (and somewhat larger) turret without a rear overhang. The modified gun cradle with bulges on the sides before the turret front is also an identifying mark. Some of the D types have cast road wheels.

Both pictures show the same tank. The changed turret shape of the D type is easy to recognize. The perforated road wheels are — as usual — without rubber pads. This meant that they made more noise. Also typical of the D type are the new-type round, armored ventilator at the back of the turret roof and the handhold on the turret sides. They allowed quicker entrance and exit.

The separation of the turret from the hull suggests an explosion caused by a direct hit to the ammunition. (1 x BA)

This photo shows clearly the new type of gun mount with the bulges on the sides. As opposed to the German 75mm gun, the Russian 7.62 cm gun had no muzzle brake.

Above: This type D has taken two direct hits on the turret. Here too, two periscopes can be seen. This photo shows clearly why the German soldiers nicknamed the Russian tanks with two round hatch covers "Mickey Mouse." (BA)

Right: A German 8.1 cm grenade launcher operates under cover of this disabled T-34/76(D). The white arrow on the turret roof was a symbol by which the Russian fighter pilots recognized their own tanks.

Upper left: As of 1943, more lettering appeared on the tanks, dedicated to the collectives or youth groups that had contributed money for the tanks. This one reads: "League of the Young Communists of Chabarovsk."

Above: An abandoned T-34/76(D) after a tank battle in the Ukraine.

Left: On this D type there has been painted on the turret, along with a number code (giving information on brigade, etc., membership), a red star and, under it, the word "Stalinetz" along with the symbol of the "Red Banner Order."

Type E, 1943

Along with many changes barely visible from outside (especially in the realm of manufacturing), it has a new commander's cupola.

Both photos show T-34/76(E) tanks put out of service in the Karosten-Schitomir area in the autumn of 1943. The commander's cupola with its two-piece cover, found only on this type, is easy to see. The turret and hull are like those of the D type.

Now the commander can get in through the commander's cupola. The E type (unlike the D) has only one "Mickey Mouse ear", that used by the gunner. The lower photo clearly shows the outside fuel tanks added to increase the range. (1 x BA)

Type F, 1943

This type lacks the commander's cupola and is recognizable only by the rounded shape of the cast turret. Only a few examples of it were built, since the T-34 with the bigger (8.5 cm) gun came out at the end of 1943 and gradually replaced the old 76 type as of the spring of 1944.

Below: This T-34/76(F) has mixed road wheels; the perforated type lack rubber pads.

Typical of the F type is the turtle-shaped cast turret with a contoured lower rim. Its lack of a commander's cupola makes it resemble the D type — which has a hexagonal turret.

The T-34/85

As an answer to the superior German Kampfpanzer V (Panther) and VI (Tiger), not only were the KV series upgraded to the Joseph Stalin (JA) version and various ever more strongly armed assault guns (SU) introduced, but the tried and true T-34 was also equipped with a higher-caliber weapon. Since the previous turret was not sufficient for the planned 8.5 cm tank gun (1943 model), the turret of the KV-85 was used. There was now room in it for a third crewman (the loader), whose work had formerly been done by the commander.

First built in the winter of 1943-44, this tank, now numbered T-34/85, appeared with the armored guard units of the Red Army in the spring of 1944. During the course of the war it was given further minor modifications, including an improved 8.5 cm gun with arrowhead shells. Thus the gun became almost the equal of the German 88mm type.

After the war (1947) a number of further — also scarcely recognizable from outside — improvements were made, so that it was now designated T-34/85 II and, as usual, its forerunner was belatedly numbered T-34/85 I. The improvements included (again) the transmission, the armor-plate structure, the fire-control equipment and the fighting-compartment ventilation.

Only in June of 1964 — after barely 25 (!) years — did production of the T-34 come to an end.

Typical of the T-34/85 is the big massive cast turret with the longer, heavier barrel of the 8.5 cm gun. Here a "German crew" poses with a captured tank. (BA)

T-34/85

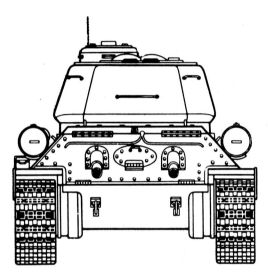

Rear View

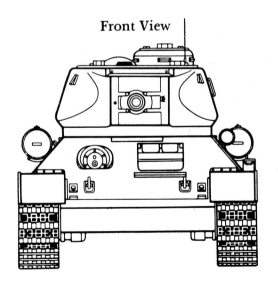

Front View

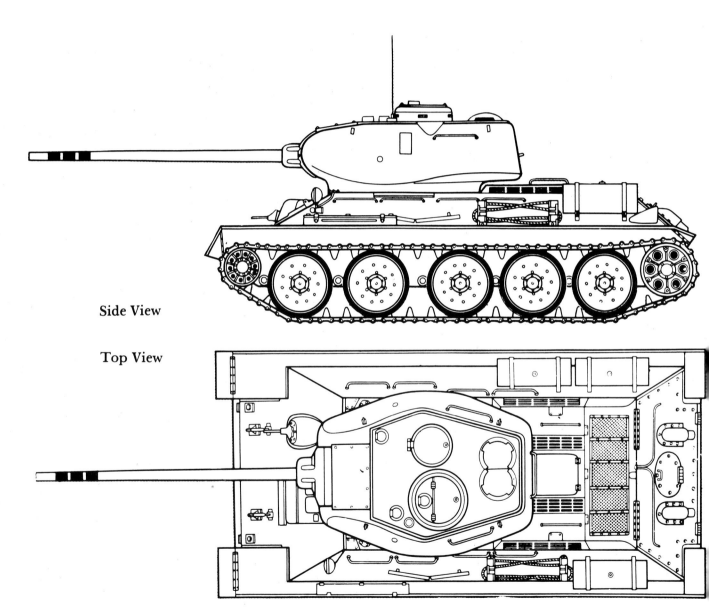

Side View

Top View

This T-34/85, taken in a Berlin suburb in April 1945, bears the inscription: "Vladimir Mayakowski" (a Russian poet).

Upper left: A T-34/85 platoon giving infantrymen a ride, as was typical then. Of particular interest are the four-digit number, the spare track links on the bow, the outside fuel tank on the apron and the inscription on the turret wall: "For Soviet Uzbekistan."

Above: A T-34/85 — photographed in 1945. The front tank is inscribed: "Forward to Berlin!"

Left: A T-34/85 with Finnish camouflage paint, as it can be seen today at the Parola Tank Museum in Finland.

This T-34/85 with the typical outside fuel tanks was captured by the 7th Panzer Division during the winter of 1944-45.

The five-man (as opposed to four in the T-34/76) crew of this T-34/85, photographed in Czechoslovakia, have utilized space by attaching all sorts of equipment to the outside of the tank.

Noteworthy features of this T-34/85 tank now in a museum are the two ventilators on the roof of the turret.

Upper left: This captured T-34/85 was used by the Wehrmacht for demonstrations. The numbers painted on it indicate armor thicknesses, number of shells and the gun's arc of elevation.

Upper left: The outside fuel tanks of this model, set at an angle to each other, are of interest.

Above: This photo clearly shows the length (L/51.5) of the gun. Before the number on the turret is a red star, which was usually applied during the last years of the war.

Left: T-34/85 and Tiger: The mixed types of road wheels are noteworthy. This rarely occurred, and almost always showed —unlike the T-34/76 — the perforated road wheels with rubber pads (as here).

Variants of the T-34

Like all dependable tank chassis, that of the T-34 was also used for other purposes on the battlefield. It served chiefly for the construction of self-propelled guns. Thus here too a self-propelled artillery vehicle — like the first assault guns made for the Wehrmacht in 1940 — was created.

It was designated SU-122 and consisted of the 12.2 cm (model 1938) field artillery howitzer on the T-34/C chassis.

SU stands for Samakhodnaya Ustanavka (= self-propelled gun). Because of its ineffective results against tanks, production of it was halted in the autumn of the same year, and its task was taken over by the SU 152 — now using the chassis of the KV tank.

Representing the escalation between tanks and antitank weapons, and showing the tendency to ever-heavier and longer-range weapons, the SU-85 appeared (even before the T-34/85 was produced!) in the autumn of 1943 as a "tank destroyer." It replaced the SU-76 (on the chassis of the T-70) which had long been used as a tank destroyer.

In the spring of 1944, the SU-100 followed as a combination of the T-34/D chassis and the 10 cm D-10 S (model 1944) field/antitank gun. This self-propelled gun was the standard weapon of the Russian antitank forces until the war ended and even beyond, until 1957, and was much feared by the Wehrmacht. It is recognizable by its strikingly long gun.

The installation of the 12.2 cm howitzer on the T-34 chassis was designated SU-122. This self-propelled gun, though, was more artillery than tank destroyer.

This picture shows the left side of the SU-122.

In comparison to the upper picture, this shows the left side of the SU-85. This self-propelled gun has clearly been turned into a tank destroyer by the long 8.5 cm gun.

An SU-85 of the 13th Polish Artillery Regiment (Sfl). The picture was taken in 1945. The white eagle of Poland can be seen ahead of the number 3 on the fighting compartment wall. The white triangle under the antenna is an identifying mark for pilots. There is probably a similar triangle on the roof.

The right front of an SU-85 with an unknown white symbol. In the background is a T-34/85. Both had the same cannon, but the SU had only a four-man crew.

Above: A practice attack of a Goliath load carrier on a captured SU-85.

Two pictures of the particularly feared SU-100. It differed from the SU-85 in having a particularly long (L/54) gun and a commander's cupola extending out somewhat over the side armor.

In addition to the already mentioned SU types, there were a great number of T-34 variants made for other purposes, such as:
— Recovery tractors
— Bridgelayers
— Mine destroyers (using rollers or explosive cord)
— Engineer tanks
— Flamethrowers. and
— Training tanks.

There were various forms of them all, sometimes built only as prototypes or in small numbers for troop testing and then rejected.

Upper right: A T-34/76(F) with a device (threshing flail) to make mines explode without damaging the tank itself.

Right: Here is a similar device attached to a T-34/85.
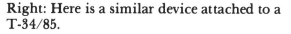

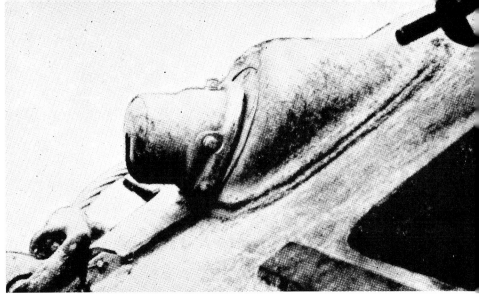

Upper left: Here a T-34/85 has a road scraper attached to its bow. Similar rigs were also based on the T-34/76.

Above: A flamethrower instead of a gun on a T-34/76(D). This flamethrowing tank was designated OT-34. It first saw service in 1944.

Left: There were various types of T-34 bridgelaying tanks. This type with the rigid "ARK" type bridge saw service in the war.

Service with Allies

When the countries of Poland, Rumania, Hungary, Bulgaria, Yugoslavia and Czechoslovakia were occupied, Communist peoples' republics were established immediately, along with fighting forces that — equipped with Russian and captured materials — took part in the last battles of the war. Here, along with those already shown, are a few more examples.

Above: A Czech crew before their T-34/76(E). On the turret is the word "Lidice."

Right: Here is a T-34/76(D) with its Polish crew. The white eagle of Poland can be seen ahead of the number on the turret.

Upper left: May 1945: A column of T-34/85 tanks of the 1st Czechoslovakian Armored Brigade marches into Prague. The three-colored Czech circle emblem can be seen next to the numbers on the turrets.

Above: Here a Polish (note the white eagle on the turret) T-34/85 of the 1st Polish Armored Brigade marches through Danzig in March of 1945.

Left: This is a T-34/85 of the 2nd Yugoslavian Armored Brigade in the streets of Trieste. The turret inscription says: "To Berlin."

T-34's in German Service

With the great numbers of destroyed and captured T-34's — especially from 1941 to 1943 — it was only natural that many units of the German Army filled their gaps with them. The tank and antitank units in particular utilized them, but so did the grenadiers, in the form of whole companies or platoons, but also as individual tanks. They all lasted as long as ammunition and spare parts were available — usually only a couple months — and always were used on the initiative of the troops. Whole units were never officially equipped with captured T-34 tanks.

Because of its "hostile" silhouette, its crews were often more or less a suicide squad in poor visibility. The German crosses were therefore painted on them considerably larger than on German tanks.

Above: In the foreground is a T-34/76(B).

Right: A T-34/76(D) with a German-installed commander's cupola. The "Op" on the hull under the turret stand for "von Oppeln-Bronikowski", the name of the regimental commander (in the spring of 1943) of Panzer Regiment 11 (6th Panzer Division), whose 7th Company was equipped completely with T-34's at that time.

A T-34/76(E) in a Rumanian city.(BA)

In front is an SU-85 with a gigantic German cross. On the bow, to the right of the gun, another one can be seen. Behind it is a T-34(E). (BA)

Post WWII

After the war ended, the T-34 — almost all in T-34/85 form — served for years in the armies of the USSR and its satellites. It was also used in numerous Asian and African countries, as well as one in Latin America.

It saw service in:

Albania, Bulgaria, East Germany, Finland, Yugoslavia, Austria, Poland, Rumania, Russia, Czechoslovakia, Hungary and Cyprus.

Afghanistan, Bangladesh, China, Iraq, Israel, Laos, Libya, Mongolia, North Yemen, North Korea, South Yemen, Syria and Vietnam.

Egypt, Algeria, Angola, Equatorial Guinea, Ethiopia, Guinea, Guinea-Bissau, Congo, Mali, Mozambique, Zambia, Somalia, Sudan, Togo, Uganda and Zimbabwe.

Cuba; also used in the wars in Korea, and Vietnam

Thus it is the most widely distributed tank in the world to date.

Both pictures show the T-34/85 in the East German army.

In the Korean War the T-34/85 was used by the armies of North Korea and the People's Republic of China. At upper left, the gun shows the results of a barrel explosion. The lettering indicates the U.S. units that put these tanks out of action.

Above: A North Vietnamese T-34/85.

Upper right: A T-34/85 captured by Hungarian rebels in 1956.

Right: An Egyptian T-34/85 captured by Israel in 1967.

Afterword

It was superior to all German tanks wherever it appeared, thanks to its low height, its strong and ideally formed armor, its long gun that could attain greater ranges as well as power and accuracy (its ballistic curve being flattened by the higher velocity), its rugged Diesel engines, its mobility on wide tracks (less ground pressure) and great power-to-weight ratio (19). Only the poor training of its crews at the time, its initial lack of radio, its relatively low numbers and poor targeting optics prevented even greater success.

The result of its appearance was the quick introduction of longer German tank guns in the Panzer III and IV, as well as the hastened development of the Tiger and Panther. The latter took on the ideal form of its armor, as did the later King Tiger. Despite the weaknesses noted above and its cramped fighting compartment, its lack of a turret ventilator (it had only a shaft to the engine room), weaknesses in its transmission and the very heavy hatch covers in the turret at first, it was one of the best tanks of past years (especially as of the C type and in T-34/85 form).

As of 1944 the units of Czechoslovakia, Poland and Yugoslavia trained in Russia were also equipped with these tanks.

In all, more than 50,000 (about 40,000 by war's end, some 12,000 afterward) of them were built (more accurate figures are not available). In 1944 and 1945 alone, 10,000 of them were built per year.

After the war, the USSR's new allies received them, as did the East German barracks Volkspolizei as of 1952. Other countries in Asia and Africa obtained them later. In the Russian army, the last of them were replaced by newer tanks only in the 1960s. They took part in the Korean War and the uprisings in East Berlin (1953) and Hungary (1956). Small African nations still used them until the late 1970s.

All in all, it was one of the best, most often built and longest-used tanks in the world!

An SU-100 of the Egyptian Army.

Left: Chinese soldiers practice close antitank combat on a T-34/85 II. This later (II) version of the T-34/85 is recognizable by the raised ventilator, now located in front of the commander's cupola.

Technical Data

	T-34/76 A	T-34/76 C	T-34/85
Crew	4	4	5
Fighting weight	26.3 tons	26.5 tons	32 tons
Overall length	5.90 m	6.75 m	8.15 m
Overall width	3.00 m	3.00 m	3.02 m
Width over tracks	2.92 m	2.92 m	2.92 m
Width mid-track	2.47 m	2.47 m	2.47 m
Overall height	2.44 m	2.62 m	2.72 m
Ground clearance	31.2 cm	31.2 cm	32.5 cm
Track extent	3.72 m	3.72 m	3.72 m
Cruising speed	40 kph	32 kph	32 kph
Fuel supply	405 l	480 l	550 l
Range on road	450 km	430 km	300 km
Range off road	200 km	200 km	200 km
Power to weight	19.0 HP/t	18.9 HP/t	15.6 HP/t
Pressure per sq.cm	0.64 kg	0.71 kg	0.81 kg
Ditch spanning	2.98 m	2.50 m	2.50 m
Vertical step	0.75 m	0.85 m	0.73 m
Climbing ability	35 [degrees]	30	30
Fording ability (without aids)	1.09 m	1.31 m	1.31 m
Engine, HP/rpm	V-2-34 V12 Diesel, 500 HP/1800 rpm		
Colling	Water	Water	Water
Transmission	Variable gear	Variable gear	Variable gear
Speeds fwd/back	4/1	4/1	5/1
Steering	Clutch+brake	Clutch+brake	Clutch+brake
Track width	47.7 cm	47.7 cm	47.7 cm
Suspension	Christie ind.	Christie ind.	Christie ind.
Primary weapon	7.62 cm L/30.5 Mod. 38 multi.	7.62 cm L/41 Mod. 40 multi.	8.5 cm M-1944 ZIS-S 53 L/51.5
Secondary weapons	2x 7.62 DT-MG	2x 7.62 DT-MG	2x 7.62 DT-MG
Shells	80	77	86
MG ammunition	2394	2394	2394
Communications	Flags	Radio	Radio
Armor: turret	15-45 mm	20-70 mm	20-75 mm
Hull side, front	45 mm	45 mm	45-47 mm
Hull side, rear	40 mm	45 mm	45 mm
Hull, front	45 mm	45-47 mm	47-34+15 mm
Hull, rear	40 mm	45 mm	45 mm
Hull bottom	15-20 mm	20 mm	18-22 mm
Hull cover	18-22 mm	18-22 mm	18-22 mm

A T-34/85 in front of the Brandenburg Gate in Berlin. It shows the strong track links and grilles (anti-hand grenade, etc.) around and over the turret.

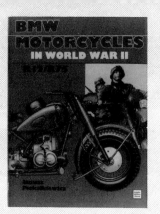

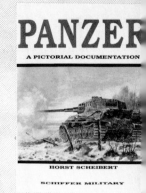